YOUR KNOWLEDGE HAS VALUE

- We will publish your bachelor's and master's thesis, essays and papers

- Your own eBook and book - sold worldwide in all relevant shops

- Earn money with each sale

Upload your text at www.GRIN.com
and publish for free

Bibliographic information published by the German National Library:

The German National Library lists this publication in the National Bibliography; detailed bibliographic data are available on the Internet at http://dnb.dnb.de .

This book is copyright material and must not be copied, reproduced, transferred, distributed, leased, licensed or publicly performed or used in any way except as specifically permitted in writing by the publishers, as allowed under the terms and conditions under which it was purchased or as strictly permitted by applicable copyright law. Any unauthorized distribution or use of this text may be a direct infringement of the author s and publisher s rights and those responsible may be liable in law accordingly.

Imprint:

Copyright © 2007 GRIN Verlag
Print and binding: Books on Demand GmbH, Norderstedt Germany
ISBN: 9783668681705

This book at GRIN:

https://www.grin.com/document/418900

Dharmendra Kumar Yadav

Anurupyena Binomial Method

An Application of Binomial Theorem in Vedic Mathematics

GRIN Verlag

GRIN - Your knowledge has value

Since its foundation in 1998, GRIN has specialized in publishing academic texts by students, college teachers and other academics as e-book and printed book. The website www.grin.com is an ideal platform for presenting term papers, final papers, scientific essays, dissertations and specialist books.

Visit us on the internet:

http://www.grin.com/

http://www.facebook.com/grincom

http://www.twitter.com/grin_com

Anurupyena Binomial Method

An Application of Binomial Theorem in Vedic Mathematics

Dharmendra Kumar Yadav

Assistant Professor and Head, Department of Mathematics
M. L. S. College, Sarisab-Pahi, Madhubani, Bihar
(Lalit Narayan Mithila University, Darbhanga)

Abstract

In the article a vedic mathematics subsutra anurupyena vidhi has been extended to find the nth power of an integer of any number of digits and of a rational number with terminating condition by applying binomial theorem for positive integral index, which is an attempt to correlate vedic mathematics with modern mathematics. The article reveals one of the logic hidden in this vedic sutra. Thus an attempt has been made to explain the unconventional aspects of method. Some people may find it difficult at first reading to understand the arithmetical operations but would play and enjoy it after some exercises.

Key Words: Sutra (Formula), Subsutra, Anurupyena Vidhi, Binomial Theorem.

Mathematics Subject Classification 2010: 00A22, 05A10, 11B65

Introduction

The Vedas are ancient Indian texts containing a record of human experience and knowledge. It is believed as the fountainhead and illimitable store-house of all knowledge. The *Vedas are Four* in number named as Yajur, Sama, Atharva and Rgveda, but they have also *Four Upavedas*: Ayurveda, Gandharvaveda, Dhanurveda, Sthapatyaveda and *Six Vedangas*. Mathematics is regarded to fall under the Upaveda Sthapatyaveda. From this, Vedic Mathematics was rediscovered by **Swami Bharati Krishna Tirthaji Maharaja** (1884-1960) of Govardhana Math, Puri during 1911 to 1918.

This system of mathematics is based on *Sixteen Sutras*: Ekadhikena Purven, Nikhilam Navatascaramam Dasatah, Urdhva-tiryagbhyam, Paravartya Yojayet, Sunyam Samayasamuccaye, Anurupye Sunyamanyat, Sankalana-vyavakalanabhyam, Puranapuranabhyam, Calana-kalanabhyam, Yavadunam, Vyastisamastih, Sesanyankena Caramena, Sopantyadvayamantyam, Ekanyunena Purvena, Gunitasamuccayah, Gunakasamuccayah and *Thirteen Sub-sutras*: Anurupyena, Sisyate Sesasamjnah, Adyamadyenantyamantyena, Kevalaih Saptakam Gunyat, Vestanam, Yavadunam Tavadunam, Yavadunam Tavadunikrtya Varganca Yojayet, Antyayordasake'pi, Antyayoreva, Samuccayagunitah, Lopanasthapanabhyam, Vilokanam, Gunitasamuccayah Samuccayagunitah; which describes the natural thinking pattern of mind. It is a common belief that through vedic mathematics we use both part of our brain keeping us mentally fit. It deals mainly with various sutras and their applications for carrying out tedious and cumbersome arithmetical operations, and to a large extent, executing them mentally in short.

As far as the binomial theorem is concerned, Greek mathematician **Euclid** knew it for exponent 2, whereas the theorem for cubes was known in India by 6th century. The binomial theorem can also be found in the work of 11th century Persian mathematician **Al-Karaji**. The binomial expansion of small degrees was known in the 13th century mathematical works of **Yang Hui** and **Chu Shih-Chieh**. In 1544, **Michael Stifel** introduced the term "binomial coefficient". **Issac Newton** is credited with the generalized binomial theorem for any rational exponent.

Preliminary Ideas

We should have the basic knowledge of the two terms one from vedic sutra and another from modern mathematics to understand the present article:

Anurupyena Sub-Formula

Swami states that the upasutra or sub-formula or sub-sutra '*Anurupyena*' means 'proportionately'. In actual application, it connotes that, in all cases where there is a rational ratio-wise relationship, the ratio should be taken into account and should lead to a proportionate multiplication or division as the case may be.

In other words, when neither the multiplicand nor the multiplier is sufficiently near a convenient power of 10 which can suitably serve as a base, we can take a convenient multiple or sub-multiple of a suitable base, as our 'working base', perform the necessary operation with its aid and then multiply or divide the result proportionately, i.e. in the same proportion as the original base may bear to the working base actually used by us. This is used in multiplying two numbers and so useful in finding square, cubes, etc.

Sharma (1996) states that this sub-formula is used to find the square of a number, which is neither near to base number nor near to sub-base number, for example 66. **Shashtri** (2011) states the limits of this upasutra that this is useful for all those numbers which are of two or three digits numbers.

Binomial Theorem
We know that if a and b are two real numbers and n is a positive integer, then

$$(a+b)^n = {}^nC_0 a^n + {}^nC_1 a^{n-1} b^1 + {}^nC_2 a^{n-2} b^2 + \ldots + {}^nC_r a^r b^{n-r} + \ldots + {}^nC_{n-1} a^1 b^{n-1} + {}^nC_n b^n$$

where ${}^nC_r = \dfrac{n!}{(n-r)!\, r!}$ and $0 \leq r \leq n$.

Anurupyena Binomial Method

Yadav (2007) has propounded this method to find the nth power of any integer of any number of digits and a rational number with terminating condition. He observed certain pattern of calculation in finding the square and cube of integers using vedic sutra but some limits have also been found in the application of this sutra. No general rule has been mentioned in the vedic mathematics text or in any others to find the nth power of any integers. He combined the anurupyena sutra and binomial theorem for positive integral index to result out this method. That's why it has been named as anurupyena binomial method. *Yadav* has discussed the following general method to find nth power:

Nth POWER OF A POSITIVE INTEGER

To find the nth power of any positive integer of any number of digits, we break the given number into two parts A starting from the left side digit and B the remaining digits of the given number. For this we first find the number of digits (let it be R) in the given number whose nth power is to be found. There may be two cases:

Case I: When the number of digits (R) is odd (let $2m+1$), then part A has $(m+1)$ digits from the left side and part B will have the remaining 'm' digits.

Case II: When the number of digits (R) is even (let $2m$), then part A has 'm' digits from the left side and part B will have the remaining 'm' digits. In this case both parts A and B have the same number of digits.

It should be kept in mind that in finding A and B, there must not be any change in the order of digits of the given number.

Now we proceed as follows:

1. Make $(n+1)$ columns and write the $(n+1)$ terms of right hand side of the binomial theorem for positive integral index

$$(A+B)^n = {}^nC_0 A^n + {}^nC_1 A^{n-1} B^1 + {}^nC_2 A^{n-2} B^2 + \ldots + {}^nC_r A^{n-r} B^r + \ldots + {}^nC_{n-1} A^1 B^{n-1} + {}^nC_n B^n$$

where $0 \leq r \leq n$, as follows:

Column−1	Column−2	Column−3	......	Column−r+1	......	Column−n	Column−n+1
${}^nC_0 A^n$	${}^nC_1 A^{n-1} B^1$	${}^nC_2 A^{n-2} B^2$	......	${}^nC_r A^{n-r} B^r$	......	${}^nC_{n-1} A^1 B^{n-1}$	${}^nC_n B^n$

We must not change the order of the columns to avoid the wrong result.

2. Evaluate each column separately.

3. Divide the number of digits R of the given number by 2 and find the quotient Q. Q may also be found as: If R = 2m+1 or 2m, where m is a positive integer, then Q = m.

4. Now to find the required result we combine all n+1 parts as:

i. Take as many digits as the quotient value is from the right side of column-(n+1) (let it be P_{n+1}) and add the remaining digits (number) in its left side in column–n.

ii. Again take as many digits as the quotient value is from the right side of the column–n (let it be P_n) and write it in the left side of P_{n+1} (like P_n P_{n+1}) and add the remaining digits (number) in its left side column-(n-1).

iii. Repeat (i) and (ii) up to column-2 till P_2 and finally find the number P_1 in column-1 after adding remaining digits (numbers) of column–2 and write P_1 to the left of P_2.

Thus we have already found the required nth power of the given positive integer as

$$P_1 P_2 P_3 \ldots P_r \ldots P_n P_{n+1}.$$

Here between any two P_i, (where i = 1, 2, ..., n+1), there is no operation like multiplication. It is just representation of numbers only. Here all P_2, P_3,............, P_{n+1} are numbers of m digits, where m is equal to the quotient Q.

From this method we can develop methods for different exponent values of n one by one as follows:

TO FIND THE SQUARE OF A POSITIVE NUMBER

To find the square of any positive integer of any number of digits, we break the given number into two parts A starting from the left side digit and B the remaining digits of the given number using the process discussed above. Thereafter we proceed as follows:

1. Make 3 columns and write the 3 terms of right hand side of the binomial theorem for positive integral index
$$(A+B)^2 = {}^2C_0 A^2 + {}^2C_1 A^1 B^1 + {}^2C_2 B^2$$
as follows:

Column – 1	Column – 2	Column – 3
A^2	$2 \times A \times B$	B^2

2. Evaluate each column separately.

3. Divide the number of digits R of the given number by 2 and find the quotient Q. Q may be found as: If R = 2m+1 or 2m, where m is a positive integer, then Q = m.

4. Now to find the required result we combine all 3 parts as:

   i. Take as many digits as the quotient value is from the right side of column–3 (let it be P_3) and add the remaining digits (number) in its left side in column–2.

   ii. Again take as many digits as the quotient value is from the right side of the column–2 (let it be P_2) and write it in the left side of P_3 (like $P_2 P_3$) and add the remaining digits (number) in its left side column-1.

   iii. Finally find the number P_1 in column-1 after adding remaining digits (numbers) of column–2 and write P_1 to the left of P_2.

Thus we find the square of the positive integer as $P_1 P_2 P_3$.

Let us examine the method with following examples:

Example 1: Find the square of 13.
The given number is of 2 digit, therefore R = 2, A = 1, B = 3. We make the three columns as:

$$
\begin{array}{ccc}
Column-1 & Column-2 & Column-3 \\
A^2 & 2AB & B^2 \\
1^2 = 1 & 2 \times 1 \times 3 = 6 & 3^2 = 9
\end{array}
$$

Thus $13^2 = 169$. Since here R = 2, Q = 1, we take only one digit from right side of column-3 & column-2 respectively.

Example 2: Find the square of 312.
Here R = 3, A = 31, B = 2. We make the three columns as:

$$
\begin{array}{ccc}
Column-1 & Column-2 & Column-3 \\
A^2 & 2AB & B^2 \\
31^2 = 961 & 2 \times 31 \times 2 = \underline{12}4 & 2^2 = 4 \\
+12 & 4 & 4 \\
\hline
973 & 4 & 4
\end{array}
$$

Thus $31^2 = 97344$. Here R = 3, Q = 1, therefore we have taken one digit from right side. The underlined number in column-2 has been added in its left side in column-1.

Example 3: Find the square of 8394.
Here R = 4, A = 83, B = 94. We make the three columns as:

$$
\begin{array}{ccc}
Column-1 & Column-2 & Column-3 \\
A^2 & 2AB & B^2 \\
83^2 = 6889 & 2 \times 83 \times 94 = 15604 & 94^2 = \underline{88}36 \\
+156 & +88 & 36 \\
\hline
7045 & \underline{15692} & 36
\end{array}
$$

Thus $8394^2 = 70459236$. Here R = 4 and Q = 2. Therefore we have taken two digits from right side.

Example 4: Find the square of 33333.

Here R = 5, A = 333, B = 33. We make the three columns as:

$$
\begin{array}{ccc}
Column-1 & Column-2 & Column-3 \\
A^2 & 2AB & B^2 \\
333^2 = 110889 & 2\times 333\times 33 = 21978 & 33^2 = \underline{10}89 \\
+219 & +10 & 89 \\
\hline
111108 & 21\underline{9}88 & 89
\end{array}
$$

Thus 33333^2 = 1111088889. Here R = 5 and Q = 2, Therefore we have taken two digits from right side.

Example 5: Find the square of 389456.

Here R = 6, A = 389, B = 456. We make the three columns as:

$$
\begin{array}{ccc}
Column-1 & Column-2 & Column-3 \\
A^2 & 2AB & B^2 \\
389^2 = 151321 & 2\times 389\times 456 = 354768 & 456^2 = \underline{207}936 \\
+354 & +207 & 936 \\
\hline
151675 & 3\underline{54}975 & 936
\end{array}
$$

Thus 389456^2 = 151675975936. Here R = 6 and Q = 3, Therefore we have taken three digits from right side.

Example 6: Find the square of 1111111111.

Here R = 10, A = 11111, B = 11111. We make the three columns as:

$$
\begin{array}{ccc}
Column-1 & Column-2 & Column-3 \\
A^2 & 2AB & B^2 \\
11111^2 = 123454321 & 2\times 11111\times 11111 = 246908642 & 11111^2 = \underline{123454}321 \\
+2469 & +1234 & 54321 \\
\hline
123456790 & 246\underline{9}09876 & 54321
\end{array}
$$

Thus 1111111111^2 = 1234567900987654321. Here R = 10 and Q = 5, therefore we have taken five digits from right side.

Example 7: Find the square of 1000.

Here R = 4, A = 10, B = 00. We make the three columns as:

$$
\begin{array}{ccc}
Column-1 & Column-2 & Column-3 \\
A^2 & 2AB & B^2 \\
10^2 = 100 & 2\times 10\times 00 = 0 & 00^2 = 0 \\
100 & 00 & 00
\end{array}
$$

Thus $(1000)^2 = 1000000$. Here R = 4 and Q = 2, therefore we have taken two digits from right side of column-3 and column-2 respectively. Since we have to take two digits from the right side of column-3 and column-2, therefore we can write 0 = 00.

Example 8: Find the square of 6.

Here the given number is of only one digit and to apply the method, we must have at least two digits. Therefore we write it as 6 = 06. Now R = 1, A = 0, B = 6. We make the three columns as:

$$
\begin{array}{ccc}
Column-1 & Column-2 & Column-3 \\
A^2 & 2AB & B^2 \\
0^2 = 0 & 2\times 0\times 6 = 0 & 6^2 = \underline{3}6 \\
0 & \underline{+3} & 6 \\
0 & 3 & 6
\end{array}
$$

Thus $(6)^2 = (06)^2 = 36$. Here A = 0 and Q = 1, therefore we have taken one digit from the right side of column-3 and column-2 respectively.

TO FIND THE THIRD POWER OF A POSITIVE INTEGER

To find the third power or cube of any positive integer of any number of digits, we break the given number into two parts A starting from the left side digit and B the remaining digits of the given number using the process discussed earlier. Thereafter we proceed as follows:

1. Make 4 columns and write the 4 terms of right hand side of the binomial theorem for positive integral index

$$(A+B)^3 = {}^3C_0 A^3 + {}^3C_1 A^2 B^1 + {}^3C_2 A^1 B^2 + {}^3C_3 B^3$$

   as follows:

Column–1	Column–2	Column–3	Column–4
A^3	$3 \times A^2 \times B$	$3 \times A \times B^2$	B^3

2. Evaluate each column separately.

3. Divide the number of digits R of the given number by 2 and find the quotient Q. Q may also be found as: If R = 2m+1 or 2m, where m is a positive integer, then Q = m.

4. Now to find the required result we combine all 4 parts as:

i. Take as many digits as the quotient value is from the right side of column-4 (let it be P_4) and add the remaining digits (number) in its left side in column–3.

ii. Again take as many digits as the quotient value is from the right side of the column–3 (let it be P_3) and write it in the left side of P_4 (like P_3 P_4) and add the remaining digits (number) in its left side column-2.

iii. Repeat (i) and (ii) up to column-2 till P_2 and finally find the number P_1 in column-1 after adding remaining digits (numbers) of column–2 and write P_1 to the left of P_2.

Thus we have already found the required nth power of the given positive integer as $P_1 P_2 P_3 P_4$.

Let us see the method working in finding the cube of the numbers as:

Example 9: Find the cube of 68.
Here the given number is of two digits, therefore R=2, A=6, B=8. Now we make the four columns as follows:

Column – 1	Column – 2	Column – 3	Column – 4
A^3	$3 \times A^2 \times B$	$3 \times A \times B^2$	B^3
$6^3 = 216$	$3 \times 6^2 \times 8 = 864$	$3 \times 6 \times 8^2 = 1152$	$8^3 = \underline{5}12$
+98	+120	+51	2
314	9\underline{8}4	1\underline{20}3	2

Thus $68^3 = 314432$. Here R=2, Q=1. Therefore we have taken one digit from right side of column-4, column-3, and column-2 respectively.

Example 10: Find the cube of 668.
Here the given number is of three digits, therefore R = 3, A = 66, B = 8. Now we make the four columns as follows:

Column – 1	Column – 2	Column – 3	Column – 4
A^3	$3 \times A^2 \times B$	$3 \times A \times B^2$	B^3
$66^3 = 287496$	$3 \times 66^2 \times 8 = 104544$	$3 \times 66 \times 8^2 = 12672$	$8^3 = \underline{5}12$
+10581	+1272	+51	2
298077	1058\underline{1}6	12\underline{72}3	2

Thus $668^3 = 298077632$. Here R = 3 and Q=1, therefore we have taken one digit from right side.

Example 11: Find the cube of 3333.
Here the given number is of four digits, therefore R = 4, A = 33, B = 33. Now we make the four columns as follows:

Column–1	Column–2	Column–3	Column–4
A^3	$3 \times A^2 \times B$	$3 \times A \times B^2$	B^3
$33^3 = 35937$	$3 \times 33^2 \times 33 = 107811$	$3 \times 33 \times 33^2 = 107811$	$33^3 = \underline{35}937$
$+1088$	$+1081$	$+359$	37
37025	10<u>8892</u>	10<u>8170</u>	37

Thus 3333^3 = 37025927037. Here R=4 and Q=2, therefore we have taken two digits from right side.

Example 12: Find the cube of 1000.
Here the given number is of four digits, therefore R = 4, A = 10, B = 00. Now we make the four columns as follows:

Column–1	Column–2	Column–3	Column–4
A^3	$3 \times A^2 \times B$	$3 \times A \times B^2$	B^3
$10^3 = 1000$	$3 \times 10^2 \times 00 = 0$	$3 \times 10 \times 00^2 = 0$	$00^3 = 0$
1000	0	0	0
1000	00	00	00

Thus $(1000)^3$ = 1000000000. Here R = 4 and Q=2, therefore we have taken two digits from right side of column-4, column-3, column-2 respectively. Since here we have to take two digits from the right side and in last three columns we have only one digit, so we can write 0=00 as it is necessary to follow this rule according to the method discussed earlier.

TO FIND THE FOURTH POWER OF A POSITIVE INTEGER

To find the 4th power of any positive integer of any number of digits, we break the given number into two parts A starting from the left side digit and B the remaining digits of the given number using the procedures discussed earlier. Thereafter we proceed as follows:

1. Make 5 columns and write the 5 terms of right hand side of the binomial theorem for positive integral index

$$(A+B)^4 = {}^4C_0 A^4 + {}^4C_1 A^3 B^1 + {}^4C_2 A^2 B^2 + {}^4C_3 A^1 B^3 + {}^4C_4 B^4$$

as follows:

Column-1	Column-2	Column-3	Column-4	Column-5
A^4	$4 \times A^3 \times B$	$6 \times A^2 \times B^2$	$4 \times A \times B^3$	B^4

2. Evaluate each column separately.

3. Divide the number of digits R of the given number by 2 and find the quotient Q. Q may also be found as: If $R = 2m+1$ or $2m$, where m is a positive integer, then $Q = m$.

4. Now to find the required result we combine all five parts as:

i. Take as many digits as the quotient value is from the right side of column-5 (let it be P_5) and add the remaining digits (number) in its left side in column–4.

ii. Again take as many digits as the quotient value is from the right side of the column–4 (let it be P_4) and write it in the left side of P_5 (like $P_4 P_5$) and add the remaining digits (number) in its left side column–3.

iii. Repeat (i) and (ii) up to column-2 till P_2 and finally find the number P_1 in column-1 after adding remaining digits (numbers) of column–2 and write P_1 to the left of P_2.

Thus we find the required 4th power of the given positive integer as $P_1 P_2 P_3 . P_4 P_5$.

Let us see the method working on some examples:
Example 13: Find the fourth power of 12.
Since the given number is of two digits, therefore R = 2, A = 1, B = 2. Now we make the five columns as follows:

$Column-1$	$Column-2$	$Column-3$	$Column-4$	$Column-5$
A^4	$4\times A^3 \times B$	$6\times A^2 \times B^2$	$4\times A\times B^3$	B^4
$1^4=1$	$4\times 1^3 \times 2 = 8$	$6\times 1^2 \times 2^2 = 24$	$4\times 1\times 2^3 = 32$	$2^4 = 16$
$+1$	$+2$	$+3$	$+1$	16
2	$\underline{1}0$	$\underline{2}7$	$\underline{3}3$	6

Thus $12^4 = 20736$. Here R = 2 and Q = 1, therefore we have taken one digit from right side.

Example 14: Find the fourth power of 111.
Here the given number is of three digits, therefore R = 3, A = 11, B = 1. Now we make the five columns as follows:

$Column-1$	$Column-2$	$Column-3$	$Column-4$	$Column-5$
A^4	$4\times A^3 \times B$	$6\times A^2 \times B^2$	$4\times A\times B^3$	B^4
$11^4 = 14641$	$4\times 11^3 \times 1 = 5324$	$6\times 11^2 \times 1^2 = 726$	$4\times 11\times 1^3 = 44$	$1^4=1$
$+539$	$+73$	$+4$	44	1
15180	$\underline{5}397$	$\underline{7}30$	4	1

Thus $111^4 = 151807041$. Here R = 3 and Q = 1, therefore we have taken one digit from right side.

Example 15: Find the fourth power of 1111.
Here the given number is of four digits, therefore R = 4, A = 11, B = 11. Now we make the five columns as follows:

$Column-1$	$Column-2$	$Column-3$	$Column-4$	$Column-5$
A^4	$4\times A^3 \times B$	$6\times A^2 \times B^2$	$4\times A\times B^3$	B^4
$11^4 = 14641$	$4\times 11^3 \times 11 = 58564$	$6\times 11^2 \times 11^2 = 87846$	$4\times 11\times 11^3 = 58564$	$11^4 = \underline{14}641$
$+594$	$+884$	$+587$	$+146$	41
15235	$\underline{594}48$	$\underline{884}33$	$\underline{587}10$	41

Thus $(1111)^4 = 1523548331041$. Here R = 4 and Q = 2, therefore we have taken two digits from right side.

Example 16: Find the fourth power of 1302.
Here R = 4, A = 13, B = 02. We make the five columns as:

$Column-1$	$Column-2$	$Column-3$	$Column-4$	$Column-5$
A^4	$4 \times A^3 \times B$	$6 \times A^2 \times B^2$	$4 \times A \times B^3$	B^4
$13^4 = 28561$	$4 \times 13^3 \times 02 = 17576$	$6 \times 13^2 \times 02^2 = 4056$	$4 \times 13 \times 02^3 = 416$	$02^4 = 16$
+176	+40	+4	4̲16	16
28737	176̲16	40̲60	16	16

Thus $(1302)^4 = 2873716601616$. Here R = 4 and Q = 2. Therefore we have taken two digits from the right side of column-5, column-4, column-3, column-2 respectively.

TO FIND THE FIFTH POWER OF A POSITIVE INTEGER

To find the 5th power of any positive integer of any number of digits, we break the given number into two parts A starting from the left side digit and B the remaining digits of the given number using the procedures discussed earlier. Thereafter we proceed as follows:

1. Make 6 columns and write the 6 terms of right hand side of the binomial theorem for positive integral index

$$(A+B)^5 = {}^5C_0 A^5 + {}^5C_1 A^4 B^1 + {}^5C_2 A^3 B^2 + {}^5C_3 A^2 B^3 + {}^5C_4 A^1 B^4 + {}^5C_5 B^5$$

as follows:

Column–1	Column–2	Column–3	Column–4	Column–5	Column–6
A^5	$5 \times A^4 \times B$	$10 \times A^3 \times B^2$	$10 \times A^2 \times B^3$	$5 \times A \times B^4$	B^5

2. Evaluate each column separately.

3. Divide the number of digits R of the given number by 2 and find the quotient Q. Q may also be found as: If $R = 2m+1$ or $2m$, where m is a positive integer, then $Q = m$.

4. Now to find the required result we combine all six parts as:

   i. Take as many digits as the quotient value is from the right side of column-6 (let it be P_6) and add the remaining digits (number) in its left side in column–5.

   ii. Again take as many digits as the quotient value is from the right side of the column–5 (let it be P_5) and write it in the left side of P_6 (like $P_5 P_6$) and add the remaining digits (number) in its left side column-4.

   iii. Repeat (i) and (ii) up to column-2 till P_2 and finally find the number P_1 in column-1 after adding remaining digits (numbers) of column–2 and write P_1 to the left of P_2.

Thus we have already found the required 5th power of the given positive integer as $P_1P_2P_3P_4P_5P_6$.

Example 17: Find the 5th power of 23.
Since the given number is of two digits therefore R = 2, A = 2, B = 3. We have six columns:

$Column-1$	$Column-2$	$Column-3$	$Column-4$	$Column-5$	$Column-6$
A^5	$5 \times A^4 \times B$	$10 \times A^3 \times B^2$	$10 \times A^2 \times B^3$	$5 \times A \times B^4$	B^5
$2^5 = 32$	$5 \times 2^4 \times 3 = 240$	$10 \times 2^3 \times 3^2 = 720$	$10 \times 2^2 \times 3^3 = 1080$	$5 \times 2 \times 3^4 = 810$	$3^5 = \underline{243}$
$+32$	$+83$	$+116$	$+83$	$+24$	3
64	323	836	1163	834	3

Thus $(23)^5 = 6436343$. Since Q = 1, we have taken only one digit from right side.

TO FIND THE SIXTH POWER OF A POSITIVE INTEGER

To find the 6th power of any positive integer of any number of digits, we break the given number into two parts A starting from the left side digit and B the remaining digits of the given number using the method discussed earlier. Thereafter we proceed as follows:

1. Make 7 columns and write the 7 terms of right hand side of the binomial theorem for positive integral index

$$(A+B)^6 = {}^6C_0 A^6 + {}^6C_1 A^5 B^1 + {}^6C_2 A^4 B^2 + {}^6C_3 A^3 B^3 + {}^6C_4 A^2 B^4 + {}^6C_5 AB^5 + {}^6C_6 B^6$$

as follows:

Column–1	Column–2	Column–3	Column–4	Column–5	Column–6	Column–7
A^6	$6 \times A^5 \times B$	$15 \times A^4 \times B^2$	$20 \times A^3 \times B^3$	$15 \times A^2 \times B^4$	$6 \times A \times B^5$	B^6

2. Evaluate each column separately.

3. Divide the number of digits R of the given number by 2 and find the quotient Q. Q may also be found as: If R = 2m+1 or 2m, where m is a positive integer, then Q = m.

4. Now to find the required result we combine all seven parts as:

i. Take as many digits as the quotient value is from the right side of column-7 (let it be P_7) and add the remaining digits (number) in its left side in column–6.

ii. Again take as many digits as the quotient value is from the right side of the column–6 (let it be P_6) and write it in the left side of P_7 (like P_6 P_7) and add the remaining digits (number) in its left side column–5.

iii. Repeat (i) and (ii) up to column-2 till P_2 and finally find the number P_1 in column-1 after adding remaining digits (numbers) of column–2 and write P_1 to the left of P_2.

Thus we find the required 6th power of the given positive integer as $P_1 P_2 P_3 . P_4 P_5 P_6 P_7$.

Example 18: Find the 6th power of 21.

Since the given number is of two digits therefore R = 2, A = 2, B = 1. Here we have to make seven columns:

Column−1	Column−2	Column−3	Column−4	Column−5	Column−6	Column−7
A^6	$6 \times A^5 \times B$	$15 \times A^4 \times B^2$	$20 \times A^3 \times B^3$	$15 \times A^2 \times B^4$	$6 \times A \times B^5$	B^6
$2^6 = 64$	$6 \times 2^5 \times 1 = 192$	$15 \times 2^4 \times 1^2 = 240$	$20 \times 2^3 \times 1^3 = 160$	$15 \times 2^2 \times 1^4 = 60$	$6 \times 2 \times 1^5 = 12$	1^6
+21	+25	+16	+6	+1	1̲2	1
85	2̲17	2̲56	1̲66	6̲1	2	1

Thus $21^6 = 85766121$. Here R = 2, Q = 1, therefore we have taken one digit from the right side of each column up to column-2.

TO FIND THE Nth POWER OF A NEGATIVE INTEGER

To find the nth power of negative integers, we use the fact $(-k)^n = (-1)^n.(k)^n$, where k is any positive integer. Now we find $(k)^n$ as the method discussed before and multiply it by +1 or –1 according to $(-1)^n$ when 'n' is even or odd. Thus we get the result.

Example 19: Find the nth power of (-21) if n=6.
Since we can write $(-21)^6 = (-1)^6 (21)^6$. But $21^6 = 85766121$ and $(-1)^6 = 1$.
Therefore $(-21)^6 = 85766121$

Example 20: Find the nth power of (-12) if n=3.
Since we can write $(-12)^3 = (-1)^3\, 12^3$. But $12^3 = 1728$ and $(-1)^3 = -1$.
Therefore $(-12)^3 = -1728$

TO FIND THE Nth POWER OF A RATIONAL NUMBER

Let the given rational number $\dfrac{P}{Q}$ be terminating and in terminating decimal form is represented by

$$\dfrac{P}{Q} = a_1 a_2 a_3 \ldots a_i . b_1 b_2 b_3 \ldots b_r$$

where '.' between a_i and b_1 is a decimal point.

To find the nth power of $\dfrac{P}{Q}$, we proceed as follows:

1. Find the nth power of $(a_1 a_2 a_3 \ldots a_i b_1 b_2 b_3 \ldots b_r)$ according to the method discussed above without considering the decimal point between a_i and b_1.

2. Now put the decimal point in the result (1) before (n.r) digits from the right side.

3. If the given rational number is negative, take positive or negative sign according to 'n' is even or odd respectively.

Thus we get the required result.

Example 21: Find the nth power of $\dfrac{155}{4}$ if n=3.

Since we have $\dfrac{155}{4} = 38.75$

Therefore to find the cube of it, we first find the cube of 3875 without considering the decimal point. Here we have R = 4, A = 38, B = 75.

Now we make the table of four columns as follows:

$Column-1$	$Column-2$	$Column-3$	$Column-4$
A^3	$3 \times A^2 \times B$	$3 \times A \times B^2$	B^3
$38^3 = 54872$	$3 \times 38^2 \times 75 = 324900$	$3 \times 38 \times 75^2 = 641250$	$75^3 = 421875$
$+3313$	$+6454$	$+4218$	421875
58185	331354	645468	75

Thus $3875^3 = 58185546875$. Since here n = 3, r = 2, therefore n.r = 3.2 = 6. Hence we put decimal point before six digits from the right side. Therefore

$$\left(\dfrac{155}{4}\right)^3 = (38.75)^3 = 58185.546875$$

Note: This method is also applicable for n = 1. But in this case no extra operation is required.

Applications

If we know the nth power of 0, 1, 2, 3, 4, 5, 6, 7, 8, 9 and have simple knowledge of multiplication then we can find the nth power of any integer or rational numbers in terminating decimal form by A-B Method.

Limits of the Method

This method is not applicable for finding the nth power of irrational numbers and rational numbers (in non-terminating decimal forms).

References

Maharaja S. B. K. T., Vedic Mathematics, Motilal Banarsi Dass Publishers Pvt Ltd, 17-22, 1992, Delhi, India

Murthy T. S. B., A Modern Introduction to Ancient Indian Mathematics, Wiley Eastern Ltd, 75-82, 1992, Delhi, India

Sharma S. S., Application of Vedic Mathematics in Competitive Arithmetic, Series-1, Sam Samyiki Ghatna Chakra, 32-33, Dec.1996, Allahabad, India

Shashtri R., Vedic Mathematics Made Easy, Arihant Publications (I) Pvt Ltd, 42-43, 51, 2011, Meerut, India

Shrivastava C. M., Vedic Ganita Paddati, Manoj Publications, 184-185, 2011, Delhi, India

Yadav D. K., Aanuruppen Binomial Method to find the nth power of Integers and Rational Numbers in Terminating Decimal Forms, Acta Ciencia Indica, 33M (2), 647-655, 2007, Meerut, India

Knnojiya D. S. & Yadav D. K., Algorithm of Aanuruppen Binomial Method, Int. J of Math. Sci. & Engg. Appls, 2(4), 101-112, 2008, Pune, India

https://en.wikipedia.org/wiki/Binomial_theorem

YOUR KNOWLEDGE HAS VALUE

- We will publish your bachelor's and master's thesis, essays and papers

- Your own eBook and book - sold worldwide in all relevant shops

- Earn money with each sale

Upload your text at www.GRIN.com
and publish for free